HOW **STEM** BUILT THE
MAYAN
EMPIRE

AMIE JANE LEAVITT

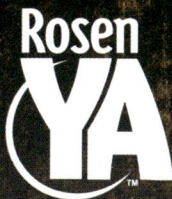

Rosen
YA™

New York

Published in 2020 by The Rosen Publishing Group, Inc.
29 East 21st Street, New York, NY 10010

First Edition

Library of Congress Cataloging-in-Publication Data

Names: Leavitt, Amie Jane, author.
Title: How STEM built the Mayan empire / Amie Jane Leavitt.
Description: First edition. | New York : Rosen Publishing, 2020. | Series: How STEM built empires | Audience: Grades: 7–12. | Includes bibliographical references and index.
Identifiers: LCCN 2019013257| ISBN 9781725341500 (library bound) | ISBN 9781725341494 (pbk.)
Subjects: LCSH: Science—Central America—History—Juvenile literature. | Technology—Central America—History—Juvenile literature. | Engineering—Central America—History—Juvenile literature. | Mathematics—Central America—History—Juvenile literature. | Mayas—Mathematics—Juvenile literature. | Mayas—History—Juvenile literature.
Classification: LCC Q127.C35 L43 2020 | DDC 509.728109/021—dc23
LC record available at https://lccn.loc.gov/2019013257

On the cover: Of all the incredible building projects undertaken by the Maya, the most iconic are their pyramid-like temples.

CONTENTS

INTRODUCTION

The Maya are an indigenous group of people who have dwelt in the Central American region for centuries. Today, there are about seven million Maya who live in the countries of Mexico, Guatemala, Belize, El Salvador, and Honduras. They speak about seventy distinct Mayan languages, and many are also bilingual in Spanish.

According to archaeological evidence, it is believed that the Mayan culture in Central America stretches back all the way to 2000 BCE. The Maya were a dominant culture in this part of the world until about 900 CE. Experts believe that the Mayan civilization reached its height somewhere between 600 CE and 900 CE. At that time, they lived in massive cities, each ruled by its own king.

Historical evidence suggests that the Maya were never part of a unified empire, but, rather, more like ancient Greece: their civilization was made up of numerous city-states, where a city was essentially its own country with its own rulers and government. This is similar to modern-day cities such as Vatican City, Singapore, and Monaco.

The Maya built these city-states in three distinct areas of Mesoamerica. The Northern Mayan Lowlands was an area situated on the Yucatán Peninsula, a body of land that juts into the Gulf of Mexico where modern-day Cancún is located. The Southern Mayan

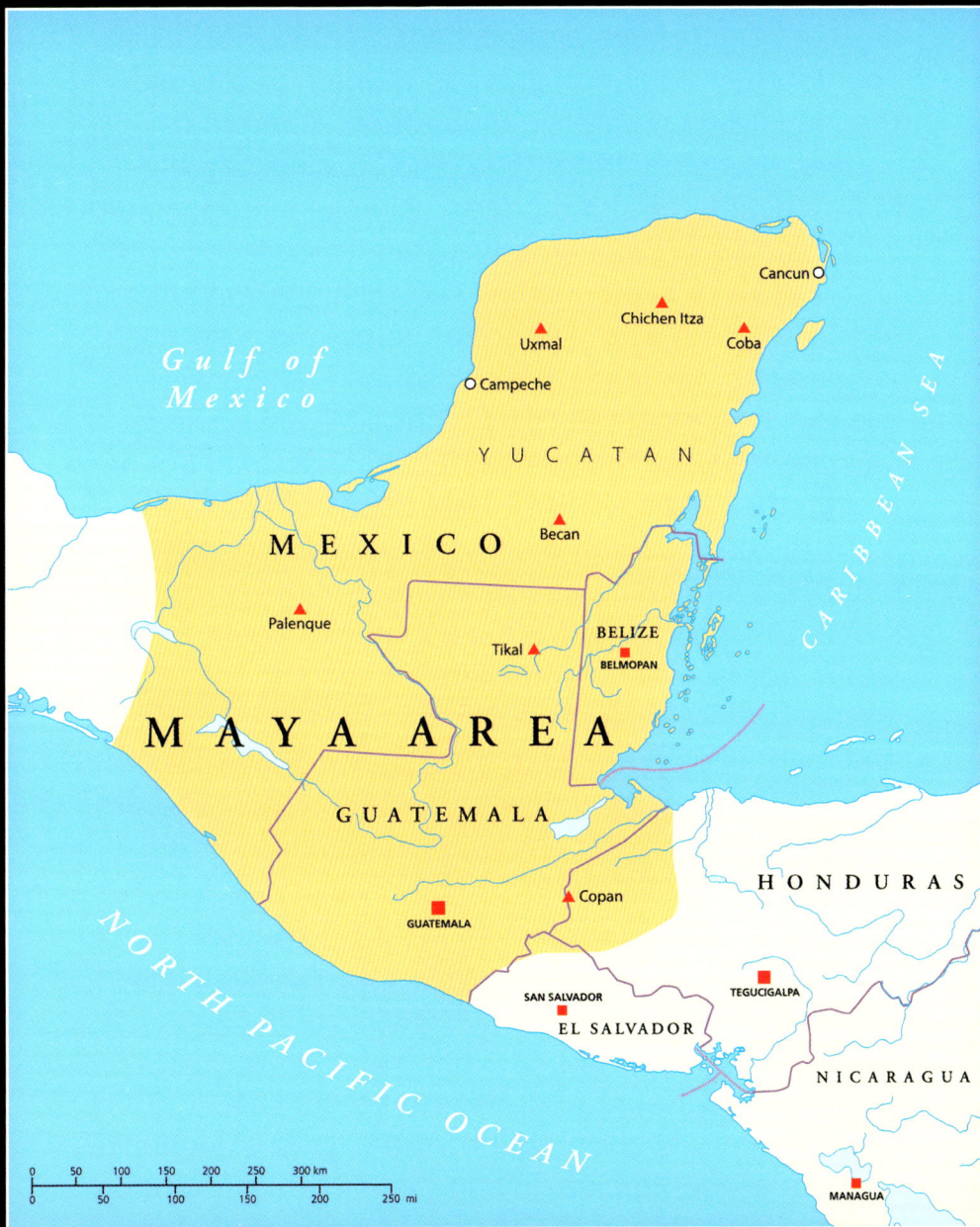

The Maya civilization stretched from sea to shining sea in Central America.

Lowlands was an area located in the Petén district of northern Guatemala and portions of modern Mexico. The Southern Mayan Highlands was an area found in the mountainous regions of southern Guatemala.

Over its nearly three thousand year history, the Maya became one of the most complex and dominant indigenous civilizations in Pre-Columbian America. At first, the Maya lived in small communities made up of simple farmers who hunted for game and grew such crops as corn (maize), beans, squash, and cassava. Eventually, the people living across Mayan society became masters in science, technology, engineering, and mathematics (STEM). They used these STEM skills to construct large urban areas with massive stepped pyramids, magnificent royal palaces, carved stone monuments and temples, and impressive public works projects, including aqueducts that fed pressurized water features in the cities, canals, walls, and road networks. It was these advancements that allowed the culture to flourish, expand, and remain powerful across such an enormous span of time.

The Mayan civilization can be divided into five distinct periods. The Preclassic (2000 BCE to 300 CE), the Classic (300 CE to 900 CE), the Postclassic (900 CE to 1550 CE), the European Colonization Era (1550 CE to 1821 CE), and the National Modern Era (1821 CE to present). Most of the STEM achievements occurred during the city construction projects of the Preclassic and Classic eras. Therefore, the achievements in STEM that occurred during those time periods will be the primary focus.

CHAPTER ONE

GREAT CITY BUILDERS

The ancient Maya were impressive city builders. There are hundreds of major city-states in Mesoamerica and likely thousands more that have yet to be excavated or are lost to time. Each city held a population of as many as fifty thousand people. This population figure included the people who lived in the city itself as well as the surrounding agricultural areas or hinterlands. Experts suspect that at its peak, the Maya culture had a population of about seven to eleven million people who were spread out across these large city-states.

PLAZAS AND PATTERNS

Based on the ruins that have already been found and excavated, it appears that nearly all Mayan cities shared certain characteristics. One similarity was the central plaza or courtyard. This structure was a long rectangular grassy area that served as the heart of every Mayan city. Archaeologists suspect that this central plaza may have been used for the Maya's great markets, public gatherings, festivals, and important religious ceremonies. These plaza markets would have been meeting places for traders from all

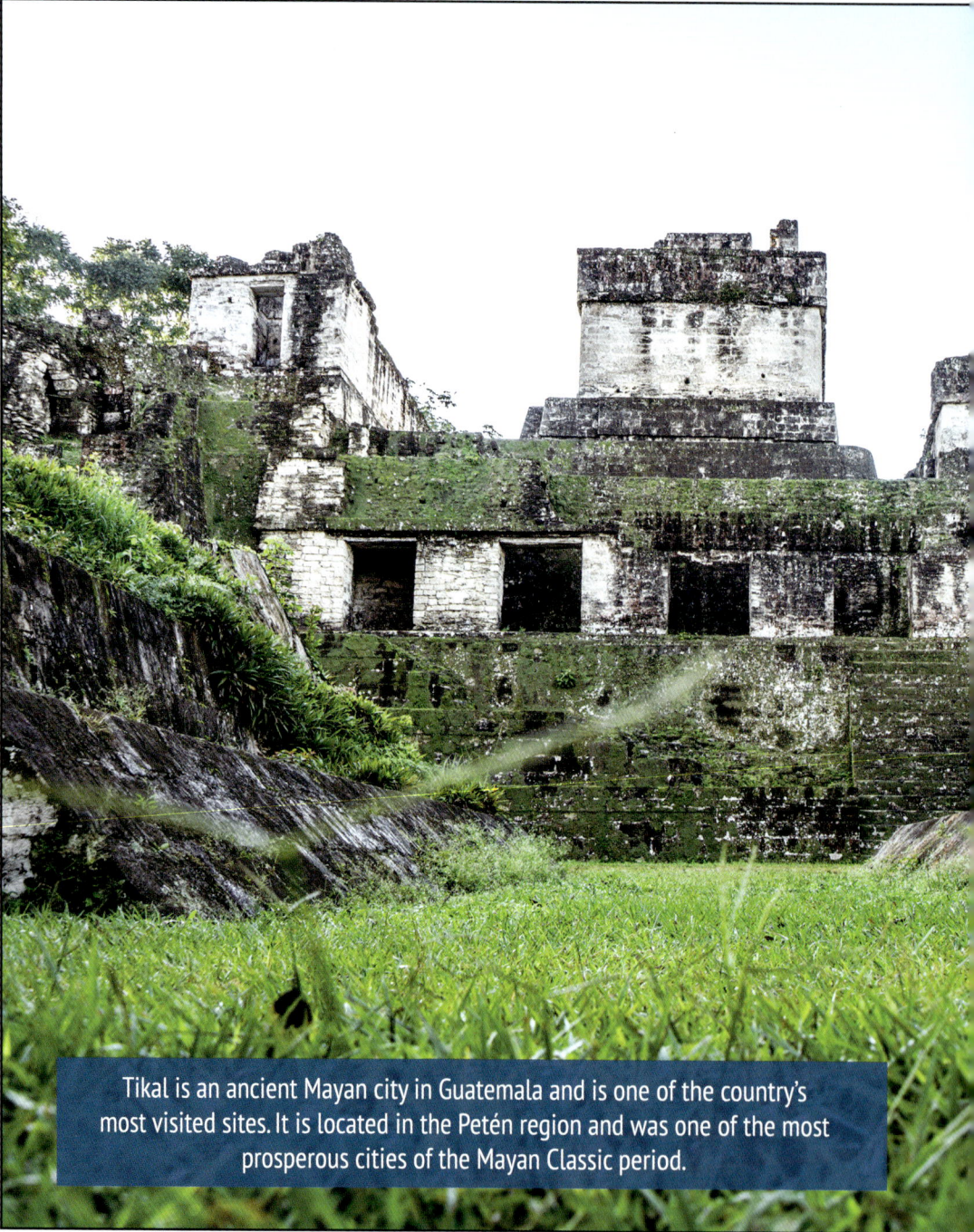

Tikal is an ancient Mayan city in Guatemala and is one of the country's most visited sites. It is located in the Petén region and was one of the most prosperous cities of the Mayan Classic period.

over—and outside—the Mayan world, as well as opportunities for residents of each city-state to make a living. Excavations of Mayan ruins show that the plazas were truly the "downtown" of the city-state. They were surrounded by the city's most important edifices, including massive stepped pyramids and stone palaces. These central areas were not only public gathering spaces, used for both commerce and recreation, they were also where the city-state's rulers and noble people dwelled.

In addition to featuring a central plaza, most Mayan cities were structured in a cluster pattern. This differs from cities built by other Mesoamerican people—the Olmecs, Toltecs, and Aztecs—who typically built their cities on a grid. At first, archaeologists theorized that the Maya just coincidentally used a more spontaneous approach to their city planning. After all, it appeared that the buildings were placed in a haphazard fashion rather than the neat and orderly straight-lined patterns of a gridded system. However, the historical understanding of Mayan cities has changed over time.

Today, most experts argue that the Maya likely used the cluster pattern for several deliberate reasons. First, they needed to build according to the topography of their areas. Most of the Mayan city-states were constructed in the geographically challenging environments of tropical rain forests with marshy swampland,

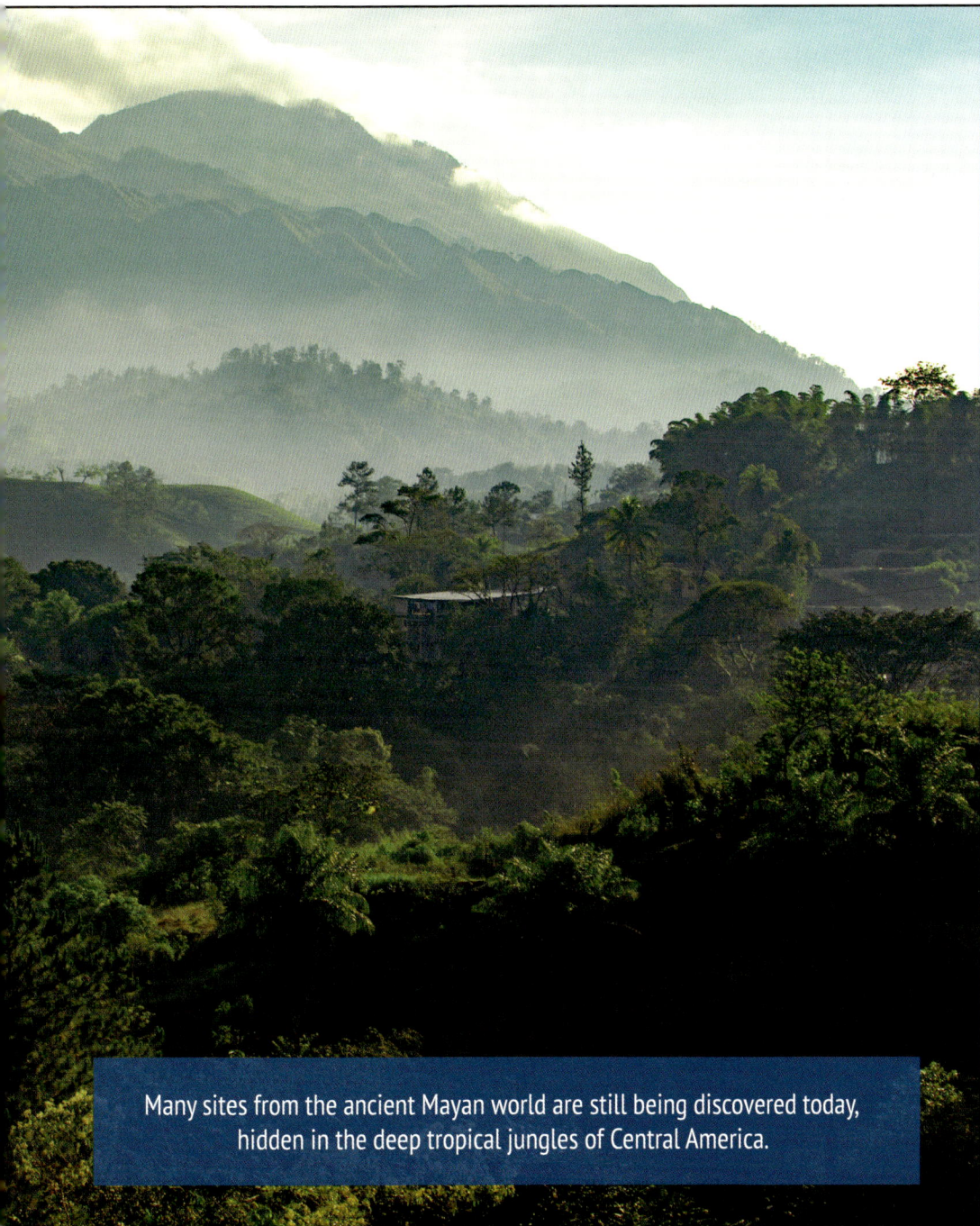

Many sites from the ancient Mayan world are still being discovered today, hidden in the deep tropical jungles of Central America.

hilly terrain, and thickly vegetated jungles. The Maya, therefore, had to work within the confines of that environment. In some cities, the people had to build terraced multilevel platforms to fit their buildings into the unyielding topography. By doing so, they were able to prevent their buildings from getting destroyed during seasons of high flooding.

Archaeologists also theorize that the Maya positioned their buildings in clustered groups to match the patterns of the sun, moon, planets, and stars. The Maya were expert astronomers and had in-depth knowledge of celestial and seasonal patterns. Some buildings were constructed so they would cast shadows in specific places at specific times of the year. Other buildings were built so Mayan astronomers could more easily view the stars and constellations.

Another theory is that the cities were actually built in a specific pattern so they would match up with Mayan hieroglyphs. When certain cities are viewed today from an aerial perspective, the building locations create patterns that look like specific glyphs. This is an intriguing theory; if it is accurate, the Maya would have needed to incorporate advanced STEM knowledge to construct a city that resembled a specific image when viewed from above. Today, getting an aerial view is relatively easy. In ancient times, to get this concept, called a bird's-eye view would have required imagining what an object looks like from a bird's perspective. To effectively plan a city using this perspective, Mayan architects and engineers would have needed to view a piece of land from some kind of elevated point—perhaps a tower or hilltop—and then sketch out how the city should be constructed to match a desired pattern. That knowledge would then need to be applied to the placement and construction of the buildings, which would not have been an easy task.

MAYAN HIEROGLYPHICS

Hieroglyphics is a form of writing that is made up of symbols, signs, and pictures. Most commonly, hieroglyphs are associated with writing used by the ancient Egyptians. However, picture writing was used by other civilizations, including the Indus, Hittites, Incans, people of Easter Island, and the Maya.

Egyptian hieroglyphics were used as early as 4000 BCE, having been found on objects from that time period. There were approximately two thousand hieroglyphic characters used by the Egyptians. Some hieroglyphs stood for a specific object, idea, or person. Others represented consonant sounds rather than concepts.

The Maya developed their hieroglyphic system of writing sometime around 300 BCE. The Maya used about eight hundred characters. Some were pictorial and represented animals, objects, and people. Others were more phonetic and represented word parts called

(continued on the next page)

(continued from the previous page)

syllables. With these characters, the Maya were able to construct complete thoughts in their writing. They used their hieroglyphics to inscribe on stelae, sculptures, sides of buildings, monuments, pottery, and on books made out of fig-bark paper with jaguar-skin covers. There are only four of these books, or codices, left in existence today: the *Dresden Codex*, *Madrid Codex*, *Paris Codex*, and *Grolier Codex*. The rest were destroyed in the 1500s by Spanish friars who feared that they contained evil practices and rituals. The four books that

Mayan hieroglyphic writing is found on many of the stelae that have been found in the region. There are hundreds of characters in the Mayan writing system.

survive contain almanacs, mathematical calculations, astronomical information, and calendars. All of this is valuable information in the history of STEM.

Experts have been working on decoding Mayan hieroglyphics for nearly a century. To date, about 85 percent of the known Mayan hieroglyphics have been decoded.

GOING AGAINST THE GRAIN

Nearly every Mayan city discovered was built in a clustered pattern. However, there is one exception: the Nixtun-Ch'ich' site in modern-day Guatemala. Researchers have used Global Positioning System (GPS) data to identify ancient buildings at the site by collecting around eighty thousand points of data across the terrain. The GPS equipment showed that—underneath the dense jungle—there were changes in elevation, which revealed the presence of buildings and streets. What surprised the researchers was that these buildings were not clustered in groups like all of the other Mayan sites. Instead, they were arranged in straight lines along east-west streets that were connected by perpendicular north-south avenues.

Timothy Pugh, an archaeologist at Queens College in New York, has been studying the site since 1995. In a Live Science article by Owen Jarus, Pugh explained certain elements featured in this city. First, the city had a main ceremonial route that ran in a straight line from east to west. About fifteen flat-topped pyramids, estimated at 100 feet (30 meters) in height, were constructed along this main corridor. Second, a triad of pyramids was built at the end of the main corridor on a raised platform. Third, from this main corridor, the city's residential areas branched off, the streets forming ninety-degree angles.

Analysis of archaeological evidence shows that Nixtun-Ch'ich' was actually a very early Mayan city. So, why did the Maya stop using this effective grid-type system and begin using their clustered patterns instead? Experts are not really sure of the answer to that question. However, there are a number of speculations. One is that the Maya may have felt too confined with this rigid layout and opted for a more open layout, which could be provided by the common cluster pattern. Another idea is that their increasing knowledge of astronomy—which was tied to Mayan religious beliefs—may have prompted the layout shift. A third idea is that this particular site was just more conducive to a grid layout because of the surrounding topography. A fourth idea is even more creative. The ancient Maya had a creation myth involving a crocodile with a hole in its back that moves from sea to land. The peninsula on which Nixtun-Ch'ich' is found resembles the shape of a crocodile crawling out of the sea. A cenote, or natural water-filled sinkhole, forms a hole on the croc's back. The city streets, in their grid pattern, form the giant reptile's scales. Even a defensive wall in the city, which cuts across the neck of the crocodile, represents another part of the Mayan myth.

Of course, all of these ideas are simply that—ideas. There is no way to know for sure why the Mayans chose to build this particular city differently from all of their others. Perhaps the answer will someday be found as further Maya sites are discovered, explored, and excavated.

PLANNING WITH A PURPOSE

Regardless of the reasons that the Maya built their cities in the clustered or gridded patterns, it is clear that they did not just throw a bunch of buildings together and call it a city. Rather, a great deal of thought and planning went into the arrangement of each city-state. These ancient people would have needed to incorporate an advanced knowledge of mathematics, astronomy, engineering, topography, geology, and archaeology when choosing where and how to build their cities.

Imagine the process of constructing a city today. How exactly would someone go about planning, designing, and building it? They would probably first take into consideration the type of land the city will be built on. Is it hilly? Are there lots of trees? Is it rocky, marshy, or an area with little water resources? Next, they would think about the types of buildings they wanted to construct and the supplies needed to build them. To stand the test of time, the construction materials would need to be strong and durable. Of course, the buildings would also need to be secure so they would not topple over. The city planners would need to consider, as well, how it could connect within itself and to surrounding areas. As a result, roads would need to be planned and built. None of these decisions could be made on a whim. To have a successful end product, numerous STEM principles work their way into city planning and design.

Constructing a city is far from an easy process, even using today's computers, high-tech gadgets, and modern building equipment. So, just think how impressive it is that the Maya built these massive cities, the ruins of which are still standing after hundreds or thousands of years, without any of these advanced modern tools. Obviously, by the look of their impressive cities, the Maya had their own methods and tools to use that were perhaps even more sophisticated in their design than those in use today. With that thought in mind, experts who study ancient Mayan ruins agree on a general consensus: only people of ingenious minds could have planned these city-states and so expertly constructed them to stand the ravages of time.

CHAPTER TWO

THE GRANDEST EDIFICES

When most people think of pyramids, they picture those found in ancient Egypt. However, the Americas actually contain more pyramids than the rest of the planet combined. The Maya were prolific pyramid builders, as were the other Mesoamerican and Incan cultures. Mayan cities were often characterized by having multiple pyramids—many grouped together in pairs or trios and crowned by stone temples.

WHY PYRAMIDS?

Pyramids are three-dimensional shapes that have a polygonal base (such as a square, pentagon, or triangle) and triangular sides. Some often wonder why ancient societies were so interested in building pyramids. Perhaps the most important reason is that the pyramid is a very stable shape—in fact, it is the most structurally stable shape that can be used for large-scale construction projects built out of stone. The pyramid allowed the ancient peoples to build tall structures that would not easily topple. Think about a pyramid in comparison to a cylindrical or rectangular pillar, tower, or box-shaped building. These can easily

Grand Mayan pyramids such as this are found throughout Central America. They each provide visual evidence of the ancient Maya's STEM skills and expertise.

be pushed over and destroyed, while a pyramid— because of its wide polygonal base and tapered triangular sides—is very difficult to demolish.

The ancient Egyptians built pyramids that have a square base and smooth triangular sides that meet at an apex at the top. Egyptian pyramids were designed as tombs, or burial places for their pharaohs. They were generally situated far away from the city center.

Mayan pyramids also have square bases, just like those of the Egyptians. However, that is where the similarities between the two cultures' pyramids end. The Maya did not build pyramids with smooth triangular sides; instead, Mayan pyramids rise up in stages, like giant stairs. Pyramidal structures with stepped sides are sometimes called ziggurats. The sides of Mayan pyramids do not form a point at the top. Rather, they are flat topped and are crowned by religious temples. The ancient Maya also considered their pyramids to be representative of mountains. They built their temples as close to the gods and the heavens as possible—and since natural mountains were available in most Mayan regions, pyramids were the next best choice.

Nearly all the known Mayan pyramids were built in the center of a city and constructed as places of worship. One known exception is the Temple of the Inscriptions in Palenque, which was built as a final resting place for Pakal, who is credited with greatly strengthening the city-state of Palenque. His tomb was discovered deep within the interior of the pyramid at the bottom of a secret staircase. While it is possible that more Mayan pyramids were also places of burial for important people, experts generally agree that the primary purpose for Mayan pyramid construction was for worship. Steep stairways on one or more sides of the pyramid led to the temple shrine at the top, which was the focal point for these structures. Not all Maya were allowed to ascend to the temples, though. Generally, that honor was restricted to the priests and kings.

CONSTRUCTION TECHNIQUES

It is widely accepted that the Maya did not have access to the wheel or beasts of burden, such as oxen, horses, donkeys, mules, camels, or llamas. Therefore, they would have needed to transport the heavy stones for their enormous construction projects from quarries using nothing but human labor. The most likely way the Maya transported these enormous stone blocks was by using logs as rollers. They would have used a lever to hoist the stones onto the rollers. Then, using ropes, they would have pushed and pulled the stones to the city center along the *sacbeob*, which was a network of paved roads. The word *sacbe* (the singular form of sacbeob) meant "white roads," a name taken from the color of the paving stones used

This primitive system shows how huge stone blocks were probably moved through the jungle to Mayan pyramid construction sites.

to create these causeways. The excavated Mayan temples throughout the region show that the people used a variety of local stone in their construction, including limestone, sandstone, and volcanic tuff.

Historical records also show that the Maya did not have access to metal tools. They, therefore, would have cut the blocks using chisels made out of black jadeite, flint, granite, quartzite, and obsidian. These extremely hard stones were readily available in the region, making them a natural choice for tool making. These same tools would have also been used to carve the intricate shapes and designs on the buildings' exteriors, which was no easy task.

Once the stones were set in place, they were held together with lime mortar. When an entire building was finished, it was covered in lime plaster. The creation of this material exhibited a surprising amount of chemical knowledge for an ancient society. To manufacture lime plaster, the Maya burned limestone using living trees, also called green wood. Using living trees instead of dead, dry wood allowed the fires to reach the extremely hot temperatures—around 1,650 degrees Fahrenheit (900 degrees Celsius)—needed to melt the limestone. It required approximately 5 tons (4.5 metric tons) of trees to produce just 1 ton (.9 metric tons) of lime plaster. To paint all of their structures and roadways with this smooth white surface coating, the Maya needed a lot of lime. This meant, in turn, chopping down massive amounts of trees around their cities to obtain the furnace wood they needed.

FAMOUS MAYAN PYRAMIDS

Though the Temple of the Inscriptions is one of the best-known Mayan pyramids—and a popular tourist destination—there were many other edifices that were equally impressive. One such structure was La Danta.

The massive stone temple known as La Danta was built around 500 BCE and is located in El Mirador, Guatemala—a city that is hidden within a dense jungle. La Danta is the primary pyramid in the city. Its bottom platform is 980 feet (298 m) wide and 2,000 feet (609 m) long. It soars some 236 feet (71.9 m) tall and has a total volume of about 99 million cubic feet (2.8 million cubic meters). These figures make La Danta one of the largest pyramids—by volume—in the world and one of the world's largest ancient structures. "We calculate that as many as 15 million man-days of labor were expended on La Danta," archaeologist Richard Hansen said in a *Smithsonian* article by Chip Brown. Hansen further explained that it would have taken twelve men to transport each block, since each weighs about 1,000 pounds (453 kilograms). The stones were cut from a quarry about 2,000 feet (610 m) away from the pyramid.

ADDING COLOR

After the plaster was applied to buildings, the monumental structures were often painted in a rainbow of colors. Many people are surprised to learn this theory; after all, the modern ruins of Mayan temples and cities are starkly white, with little trace of color anywhere. However, this is because the paint the Maya used has simply worn off over time. One of the main purposes of plastering structures with such a labor-intensive material, in fact, was to create a canvas on which the builders could paint. Inside the structures, Mayan artists often painted grand murals that depicted important events in the city-state and the reign of the kings.

Ancient peoples from around the world used their knowledge of science and the natural world to obtain the colors needed to paint their various architectural structures and art projects. The Maya were no different. It is believed that Mayan chemists used various plants, minerals, soils, and seashell pigments to create a wide spectrum of color options. They did so by using a number of methods, many of which are still a mystery to modern scientists.

However, in 2008, scientists Gary Feinman and Dean E. Arnold of the Field Museum were able to unlock the details of how the Maya created one of their famous fade-resistant pigments, dubbed Maya Blue. They examined pottery fragments under an electron microscope and found that three ingredients—extracts from indigo plant leaves, clay mineral palygorskite, and copal incense resin—were combined together to make the paint. The copal resin was used to bind everything together and give the paint its fade-resident properties. "My guess is that they probably had a large fire and a vessel over that fire where they were combining the key ingredients," Fenman explained in an article by Clara Moskowitz for Live Science. "And then they probably took pieces of the hot copal and put them in the vessel."

Color was particularly important to the ancient Maya. Some of the most common colors used in their architectural painting and art include red, blue, white, yellow, and black. In art, people were painted red, deities were painted blue, and plants and animals were painted white.

SACBE: THE "WHITE ROAD"

In addition to building monuments upward, Mayan architects and planners also built outward. Many Mayan rulers built grand highway networks to connect the buildings within their cities and their neighboring city-states. These ancient roadways were called sacbeob. These elevated roads, or causeways, were built about 10 feet (3 m) above the ground. The high water levels and marshy land made building a roadway directly on the ground impossible. To construct the roads, the Maya piled layers of stones on top of each other to reach the desired height. Then, they filled in the rocks on the road's flat surface with lime to close up any gaps. Once that was done, they painted the road with a milky, alabaster lime plaster. This

The White Road crisscrossed the emerald green rain forest, creating a highway between the major Mayan city states.

gave the roads a distinctive white color, hence the name sacbeob.

To further protect the roads in areas where flooding was common, the Maya built culverts underneath the roads. These were pipes that allowed the water to flow underneath the roadway instead of rising up and flooding the surface; they worked similarly to a modern storm drain. The Maya also built walls along the sacbeob in these areas, which further protected the roads from flooding during periods of extreme rainfall.

The sacbeob varied in size. Some were only about 9 feet (2.7 m) wide. Some others have been estimated to be up to 130 feet (39.6 m) wide. For the ancient world, this would have been absolutely massive. A lane on the US Interstate Highway System is 12 feet (3.7 m) wide. That means some Mayan roads—constructed without modern tools or knowledge—may have been wider than a huge ten-lane interstate, which would be about 120 feet (36.6 m) wide. These causeways connected cities that were hundreds of miles apart. Experts are intrigued by the fact that many of these sacbeob are built in long, straight lines that can stretch out for 60 miles (96.6 kilometers) through the forest. This raises a question: what tools did the ancient Maya use to build such impressive roadways? Perhaps with more research, this and other questions can be answered.

Modern historians and archaeologists have discovered more than 100 miles (161 km) of roadways across the Mayan sphere of influence, the oldest of which dates back to 600 BCE. The presence of such an extensive road network suggests that the Maya had regular contact with neighboring city-states. This would account for the cultural exchange between the peoples and why the design of their cities was likely so similar.

CHAPTER THREE

WATERWORKS

Over their long history in Mesoamerica, the Maya experienced a variety of challenges that were caused by either too much water or not enough water. Each city-state worked out its own individual problems through innovative engineering projects. Some built aboveground and belowground aqueduct systems. Others devised ways to conserve and treat rainwater. Still other groups built dams, reservoirs, and advanced sand filtration systems.

AQUEDUCT CONSTRUCTION

Palenque, one of the Maya's best-preserved cities, is located in southern Mexico. This city was inhabited for about eight hundred years, from about 100 CE until 900 CE. Palenque had an abundance of water within its boundaries and was also surrounded by steep mountains that sent fast-moving streams through Palenque's flat topography during rainy seasons. Kirk French, an archaeologist from Penn State University, explained in an article by Charles Q. Choi on Live Science that "the ancient Maya called this city Lakamha' or 'Big Water' because of its nine perennial waterways, 56 springs, and hundreds of meters of cascades."

An abundance of water is a great thing, especially when it comes to supporting large urban populations. However, the water cannot be allowed to just run amok—it needs to be harnessed and controlled for the benefit of the community it supports. After intense rains, streams can flood into structures, destroy farmlands, and erode natural landscapes if not monitored properly. Modern civil engineers know this principle, and there is evidence Mayan engineers did as well.

In Palenque, archaeologists have discovered that the Maya constructed sophisticated aqueduct systems as a way to harness their water supply. One of these aqueducts, the Otulum aqueduct, was constructed this way: when the mountain streams neared the city, the engineers redirected the water down a stone-lined artificial channel that was about 6 feet (2 m) deep, 10 feet (3 m) tall, 5 feet (1.5 m) wide, and 410 feet (125 m) long. At the end of the length, the channel reached a corbel arch, which marked the underground portion of the culvert. By sending the aqueduct underground, the Maya achieved two main goals. First, they were able to prevent flooding and erosion in their city. Second, they were able to preserve more livable land by diverting the streams underneath and building plazas and structures on top. Palenque was situated in an area with scarce buildable land, so creating more land by sending the streams in pipes underground allowed the population of the city to grow.

The Maya had too much water in some areas and not enough in others. They utilized their knowledge of canal systems to redirect this precious natural resource to where it was needed the most.

WATER PRESSURE

Archaeologists have found numerous aqueducts running under Palenque. One, the Piedras Bolas, is particularly interesting because the Mayas incorporated a rather advanced water pressure system within its length. This piped aqueduct was built on steep terrain that had about a 200-foot (60 m) drop in elevation in just 20 feet (6 m) of distance. The rectangular pipe at the top of this hill was 10 square feet (3 square meters) in size. At the bottom, it decreased drastically in size and was only 0.5 square feet (0.15 sq m). The difference in size means the Mayan engineers understood that water running through this pipe—aided by gravity—would build up a lot of pressure by the time it was forced through the smaller bottom opening. As gravity naturally forced water down the slope, it increased in speed. As the speed of the water increased, it was also being forced through smaller and smaller pipes. As a result, the water pressure increased enormously. By the time the water would have reached an opening at the bottom of the pipe, experts estimate that it could have shot some 20 feet (6 m) into the air.

Though there is little question the Maya knew what they were doing when it came to generating water pressure, scholars are still not quite sure what they did with it. Some suspect it may have been used for public water features, such as grand fountains or even for toilets in the royal palaces. There is evidence in Palenque of several interesting features related to water pressure inside the ruins of the royal palace, including six latrines and even sweat baths. Experts believe that these amenities would have rivaled the luxurious Roman baths. French claimed another use for water pressure in Choi's Live Science article:

> I actually think that the creation of water pressure at Palenque was a sign of wealth. It was definitely not necessary. They had water everywhere … Water pressure technology would have been useful through the display of power and knowledge, similar to how priests and shamans used astronomical events.

CORBEL ARCHES

The corbel arch was used in many places in Mayan architecture, including royal palaces, bridges, corridors, and entranceways into underground aqueducts. Both decorative and functional, the corbel arch was used to create an opening and distributed weight across the top arch. To build the arch, they started off with two rectangular stones equal in size placed on either side of an opening or gap. Then, they stacked three more rows of the same size stones on top of the first row. On the fifth, sixth, and seventh rows, they would have placed rows of slightly larger stones to form a V shape

(continued on page 35)

The corbel arch, as shown here, was used in many early Mayan structures. This architectural feature was also used by other ancient peoples across the world.

(continued from page 33)

or inverted staircase. Finally, on the last row, one long piece of stone was placed on top to span the entire distance and complete the arch. In Mayan architecture, multiple corbel arches were built in a series to form a corbel vault. These vaults were used to support the superstructure of a building's roof.

CONSERVING AND TREATING RAINWATER

In some areas of the Mayan world, the people had little to no access to fresh water. There were no streams, rivers, lakes, or springs. The only way they could get fresh water was by collecting rainwater. The Maya who lived in small villages in the area that is now northern Belize found themselves in this situation. For about half of the year, they had plenty of rainfall and could use the water that fell during each storm. However, during the other half of the year, the dry season, they would have absolutely no water to use for drinking, cooking, cleaning, growing crops, or caring for their animals. They needed to figure out a way to save it over the long term.

Their solution was to dig deep depressions in the earth and then coat them with a layer of clay to make them impermeable to water. Water would collect in these basins during the rainy seasons. Then, modern scientists believe, the Maya used a natural purification system to keep the water clean.

Research has revealed that Mayan engineers and botanists planted and regulated the kind of plant life located around these basins. They searched out and identified vegetation that would purify the water by their very presence. This natural filtration system meant the water would remain potable throughout the many months for which it needed to be stored. Archaeological finds in remote areas of the Mayan world show that even the common farmer had some level of scientific knowledge and used it to survive in difficult environments. The Maya applied their knowledge of soils when they lined the basins with clay since they knew that other types of soils would allow the water to seep into the earth. They also used

their knowledge of botany to create natural filtration systems to keep the water safe to drink.

SAVING ON A RAINY DAY

Tikal was a city in northern Guatemala, occupied by as many as eighty thousand Maya from about 300 CE and 850 CE. This city—with its impressive pyramids and royal ruins—still intrigues scientists, archaeologists, and tourists today. Not only are there impressive architectural structures in Tikal, but there is also evidence of advanced water works systems, too.

The Maya of Tikal built reservoirs at various levels throughout their city. When they cut out limestone

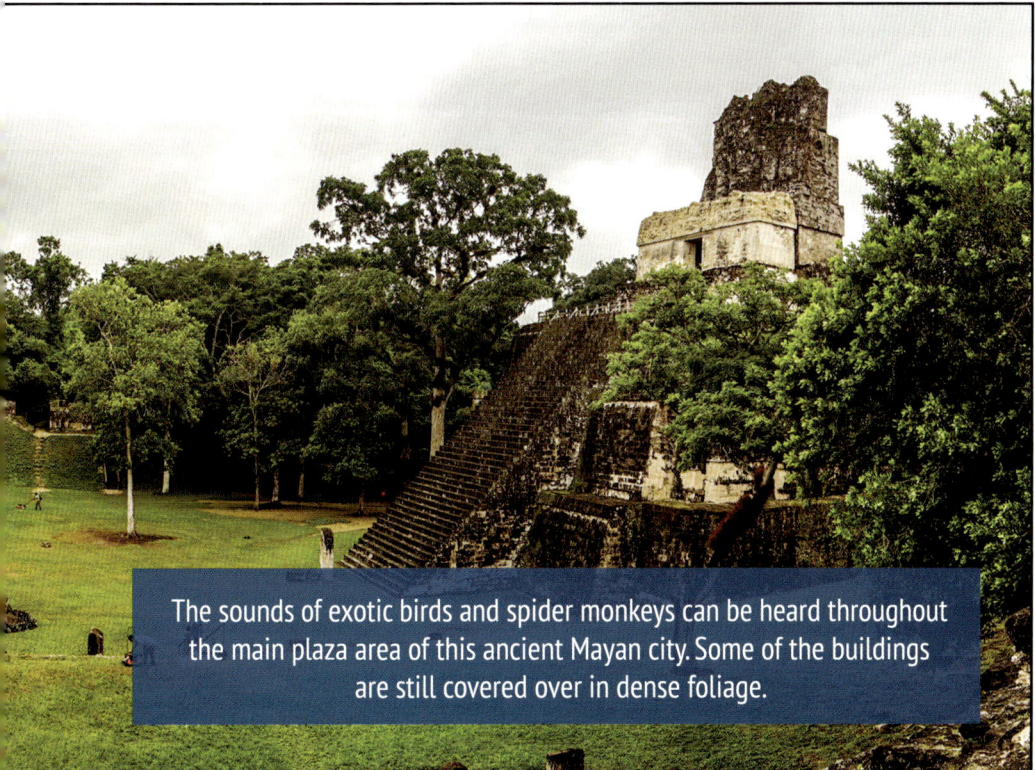

The sounds of exotic birds and spider monkeys can be heard throughout the main plaza area of this ancient Mayan city. Some of the buildings are still covered over in dense foliage.

blocks for their temples, holes were left in the ground where the rocks were quarried. These depressions created perfect reservoirs for holding water. The Maya also built dams to hold back the water in these reservoirs. The largest dam in Tikal—which is also the largest-known dam in Central America—is about 33 feet (10 m) high, 260 feet (79 m) long, and would hold about 20 million gallons (76 million liters) of water. The top of the dam was used as a roadway so that people could pass from one part of the city to the other.

The people of Tikal needed to save every drop of precipitation that fell during the rainy season if they were going to have enough water during the dry months. The very construction of the city helped contribute to their ability to save and store water. For example, city streets and buildings were paved with smooth lime plaster that prevented the water from seeping into the earth. Instead, the water would flow off into gutters and canals. The gutters and canals, which were constructed out of stone and also paved with lime plaster, extended all the way to the reservoirs.

Before the water emptied into the reservoirs, though, it had to first pass through a filtration system to clean it. The system worked like this: a stone-boxed area was filled with sand. As the water passed through the sand, impurities were filtered out. Then, the clean water emptied into the reservoir. Reservoirs that were not used for human consumption, but rather for agriculture, did not have these sand boxes attached to them. In another display of advanced STEM knowledge, the Maya in Tikal knew they must use quartz sand for their filtration system, otherwise the water may not be purified—but quartz sand is not found very close to the city-state. Scholars believe the Maya knew regular sand would not work, and so they traveled fairly long distances—some estimates are up to 20 miles (30 km)—to import sand that would more effectively purify the drinking water.

CHAPTER FOUR

MATH WHIZZES

With one look at the cities that the Maya left behind, it is obvious that they had an advanced knowledge of mathematics. There is no way they could have designed cities with such precision, constructed perfectly symmetrical structures, or figured out the proper angles necessary for water flow without the use of accurate mathematical calculation.

BASE-TWENTY SYSTEM

After decoding Mayan hieroglyphics and studying ancient codices and stelae, experts have determined that the Maya used a vigesimal—or base-twenty—number system. This means that instead of having digits from zero to nine, they had digits from zero through twenty. It also means that the place values were not ones, tens, hundreds, thousands, etc., like a base-ten system has. Instead, the place values were ones, twenties, four-hundreds, eight-thousands, etc. This vigesimal system was not unique to the Maya; many of the other Mesoamerican cultures also used a base-twenty system, including the Olmecs and the Aztecs. It is suspected that they used this system

0   1   2   3   4

5   6   7   8   9

10   11   12   13   14

15   16   17   18   19

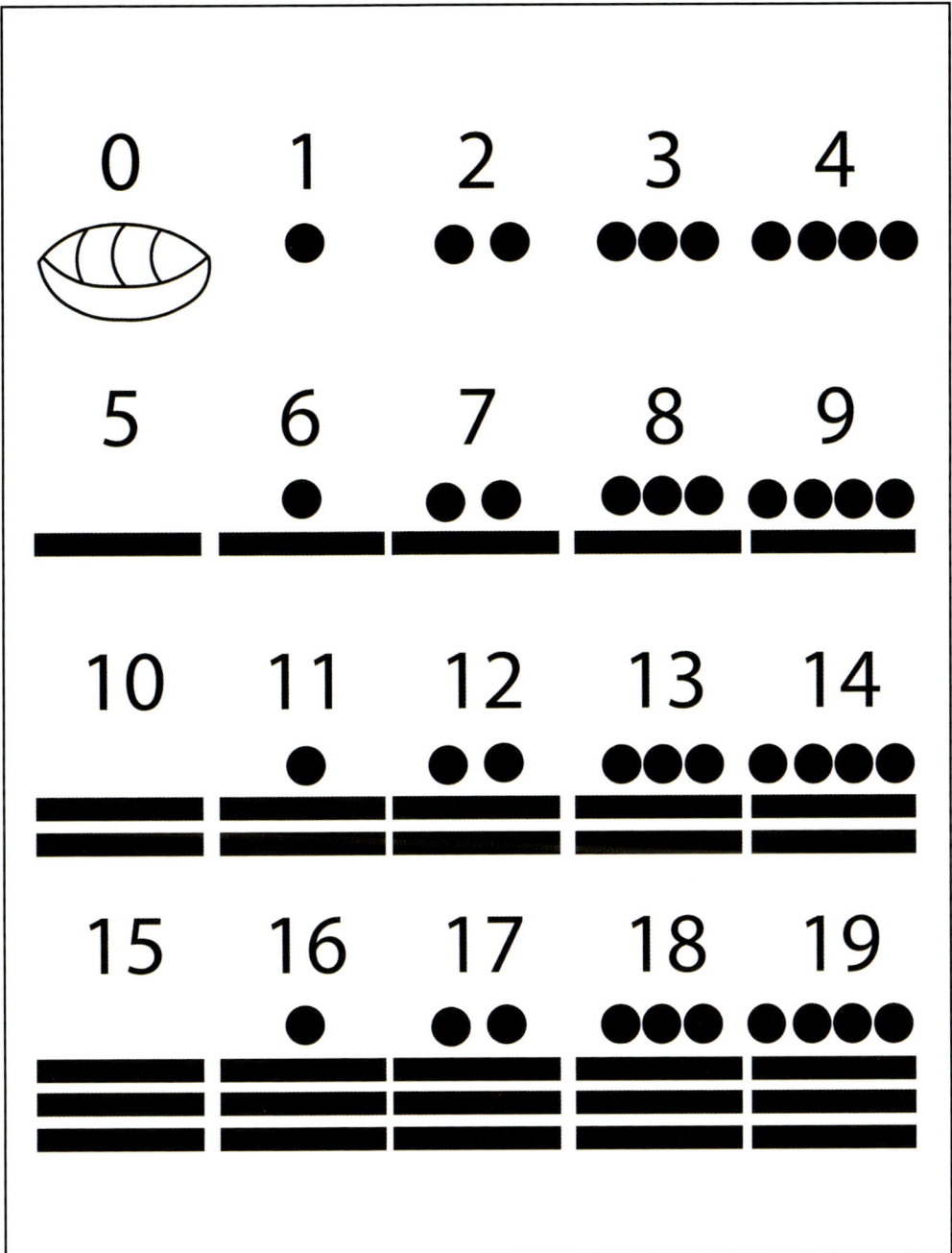

The Maya's counting system, based on twenty digits, allowed them to perform simple and complex mathematical operations.

because it utilized all twenty of the body's digits (fingers and toes) as the earliest method of counting.

Three main symbols were used in the Mayan number system: a dot representing the number one, a bar representing the number five, and a shell representing either twenty or zero. By arranging these three symbols into different positions, they were able to multiply and reach numbers into the millions.

NOT FOR NOTHING

The inclusion of zero is extremely important in any number system. It allows mathematicians to perform calculations ranging from simple arithmetic to complicated calculus. Only one other known world civilization (the Gupta Empire in India) was using the zero in its number system at the same time as the Maya. William Fash, an archaeologist from Harvard University, explained in History's *The Maya: Death of an Empire* that "the Greeks and Romans were tremendous engineers ... but were very limited by their mathematical system because they didn't have a zero ... they were able to produce great public works, philosophy, and whatnot but were really pretty lousy mathematicians compared to the Maya."

The Maya viewed zero as completion. They used zero in several ways. For one, they included zero days and zero years in their calendars. For another, they used it as a placeholder to distinguish between different types of numbers, such as 15 and 150 or 24 and 204.

THE MAYAN CALENDAR

The Maya's advanced mathematics, coupled with a wide knowledge of astronomy, helped them construct one of the most precise calendars ever created by humankind. Their calendar system was actually made up of three interlocking calendars: a solar calendar, a sacred calendar, and a long count calendar.

(continued on the next page)

Experts are still learning about the significance of the Mayan calendar system and how it allowed them to keep track of time, important events, and astronomical occurrences.

(continued from the previous page)

The solar calendar marked a solar year, which was roughly 365 days. This calendar had 18 months of 20 days each, which totaled 360 days. The final five days of the year were known as Uayeb and considered unlucky days, which prompted the Maya to devote Uayeb days to worshiping their deities.

The Mayan ceremonial or sacred calendar was made up of 260 days. This calendar marked important ceremonies and religious celebrations. It was made up of 20 periods of 13 days each.

Every fifty-two years, the solar and sacred calendar would line up and begin on the same day. This day was marked with grand celebrations since it marked what the Maya called the Calendar Round.

The long count calendar was used exactly as its name implied—it counted out long periods of time. It starts at what the Maya believed was the beginning of the world: August 11, 3114 BCE. It counted all days since that time using a 360-day year. The dates on the long calendar are written with five digits: a day (*kin*), month (*uinal*), year (*tun*), twenty-years (*katun*), and twenty-katuns (*baktun*).

GOLDEN TIME

Mathematicians theorize that the Maya tried to incorporate the mathematical relationships they observed in nature into their art and architecture.

One such relationship was the golden ratio, which is also known as the golden mean, golden section, and divine proportion. This is a special number that is equal to approximately 1.618. This ratio is thought to be aesthetically beautiful because it can be used to make golden rectangles, golden spirals, and other mathematically pleasing shapes. It is found in many places in nature, including flowers, seashells, fruits, vegetables, the human body, and honeybee colonies. The Maya are not the only civilization to utilize this mathematical ratio in their art and architecture—it is commonly found in ancient Greek, Roman, Egyptian, and Islamic masterpieces as well.

About 1.2 million people visit Chichén Itzá every year and climb the steps to the top of this most impressive Mayan temple.

Evidence of the Maya's use of the golden mean can be found in several places in their architecture. One of the most noteworthy is the pyramid El Castillo at Chichén Itzá. It was also used in the construction of the Temple of the Cross in Palenque. In the rear chambers of this building, the ratio of 1 to 1.618 can be found when comparing the building's rectangular features. These repeated squares form natural proportions and a beautifully calculated floor plan. Some experts also speculate that the golden ratio could have been used in the development of the Mayan calendar.

Interestingly, it is believed that the Mayans constructed the golden rectangles of their architecture using nothing more than sticks and strings. They would have started out by marking the corner of a square with one stick. Then, they stretched the measuring string in four different directions to mark out a perfect square with four equal sides. Next, they used half the cord to find the midway point of one side of the square. They connected this midpoint diagonally to the opposite corner of the square to make a right triangle. They then swung the string from that corner—like a compass in an arc—to form the bottom corner of the rectangle. They would have extended the sides of their original square to form the golden rectangle. This process would have been repeated and repeated to form the golden ratio pattern in the buildings.

CHAPTER FIVE

MASTERS IN ASTRONOMY

With a highly developed understanding of math and a relative mastery of the earth, the Maya understandably started looking toward the stars—and they were astronomical masters. They spent centuries tracking the sun, moon, stars, and planets. They knew how to predict solar and lunar eclipses, and they used astrological cycles to aid in the planting and harvesting of their crops.

GATHERING DATA

The Maya used a variety of tools to observe the heavens and calculate the movement of heavenly bodies. One such tool that they likely used was made up of two crossed bars. As Mayan astronomers angled the bars at the sky and focused on whatever was found at the bars' intersection, they could focus in on specific heavenly bodies. This tool would have helped them track the pathway of these objects over days, months, years, and centuries.

The Maya also used a tool called a vertical zenith sighting tube. This was a vertical opening in an underground structure that allows sunlight to form a perfect perpendicular shaft of light when the sun

crossed directly overhead twice a year. Observing the zenith is something unique to people who live between the Tropic of Cancer and the Tropic of Capricorn—regions bordering the equator—since nowhere else on earth does the sun cross at exactly 90 degrees overhead.

By making and recording these heavenly observations, the Maya were able to accurately record the passing of time. They paid attention to the spring and fall equinoxes—the times twice a year when the length of day and night is exactly twelve hours. They recorded when the zeniths occurred. They made note of the dates of summer and winter solstices. They paid attention to the shadows made at certain times of the day and year and took note of the position of the sun during sunrise and sunset as the days of the year progressed. They recorded lunar phases and observed the path of the planet Venus across the sky. All of this data was used to create one of the most sophisticated calendar systems the world has ever known.

OBSERVATORIES

Many places in the Mayan world were covered by dense jungle. Therefore, to get a good view of the sky, the ancient people had to find a way to climb above the canopy. They could scale to the top of their enormous pyramids, or they could construct special viewing structures called observatories.

One of the most famous observatories is El Caracol at Chichén Itzá. This observatory is a cylindrical tower

This rounded structure indicates that the building was used as an observatory. The windows were placed at precise locations to view astronomical occurrences.

on a square base. Its name means "the snail," given by the Spanish because of the spiral staircase that coiled around inside the tower, just like a snail's shell.

A portion of the tower's rounded dome has already crumbled. What is left reveals clues as to how the building was likely used. There are three narrow slots in the cupola that scholars believe were used to view the path of Venus. In fact, every eight years, modern astronomers have found Venus at a particular location on the horizon using one of the slotted windows. Another window was likely used to view the equinox sunsets. The building faces 27.5 degrees north of due west, which is in line with the northern path of Venus.

Plotting the path of Venus was particularly important to the Maya. They timed many of their important religious events based on their Venus observations.

THE *DRESDEN CODEX* AND ASTRONOMY

The *Dresden Codex* is one of the four massive codices that survived when Spanish friars destroyed the Mayan libraries in the sixteenth century. This codex is of particular significance to modern astronomers because it contains evidence that the Maya were able to precisely predict the dates of lunar and solar eclipses. It also shows how the Maya tracked the path of Venus across the heavens. The *Dresden Codex* contains a Venus table, which describes precise details about the planet's movements. Anthropologist Gerardo Aldana explained in an article by The Current's Jim Logan that these calculations seem to be the work of one individual Mayan astronomer. "This person, who's witnessing events at this one city [Chichén Itzá] during this very specific period of time, created, through their own creativity, this mathematical innovation."

The book also has dozens of almanacs that detail what daily life was like for the ancient Maya. The 365-day and 260-day calendars are used as forms of timekeeping to describe daily events related to agriculture, weather, worship schedules, and commerce. The priests who kept record of the days

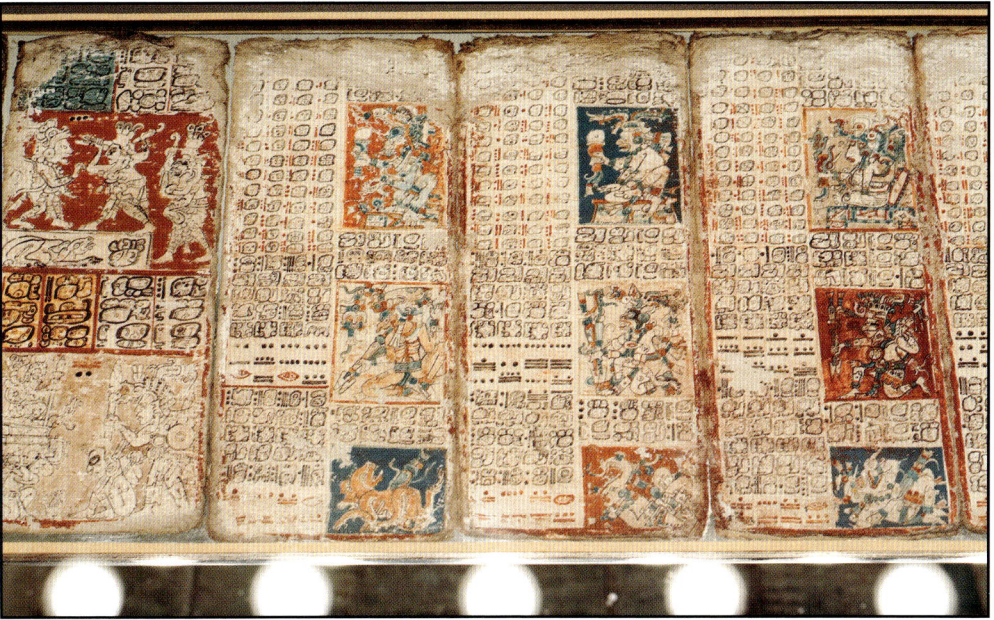

Few Mayan codices are left, and those that remain are cherished. This manuscript is the *Dresden Codex*, which is displayed under glass at a museum in Dresden, Germany.

used this information—which was passed down from generation to generation—to make predictions about future events. They also continually added to their codices by using information gathered from observatories, shadow-casting devices, and observations of the horizons.

The *Dresden Codex* is named after the city in Germany (Dresden) where it has been housed since the 1700s. The *Dresden Codex* was significantly damaged during the bombing of the city during World War II. Fortunately, facsimiles of the book were made prior to that time and stored in another location so the book can still be studied and analyzed today.

MAYAN BALL GAMES

Mayan cities have many features in common. One is the ball court. This sporting arena is also common to other Mesoamerican cities as well—in fact, there are an estimated 1,300 ball courts in this region that have been excavated. The games played on these courts were a focal point of Mayan life, and experts believe it showed each city's wealth and power. Chichén Itzá has the largest-known court in the Mayan world.

This game, which the Mayans called *pok-ta-pok*, was played by hitting a rubber ball with the elbows, knees,

On the wall, toward the left in this image, a ring can be seen; this ring would have been used for the Maya's ball game. This court is located at the Temple of the Jaguars in Chichén Itzá.

torsos, and hips—no hands allowed. The object was to get the ball through a stone ring that was attached about 20 feet (6 m) high on the stone walls of the court.

These Mesoamerican ball games illustrate this region's knowledge of STEM principles. First, the game was played with rubber balls, demonstrating an advanced chemical knowledge of the chemical applications for natural latex. Second, the game was directly tied to the Mayan knowledge of astronomy. The game represented movement of the sun, moon, and planets in the sky. When excavating ball court ruins in Mexico in 2012, archaeologists found that the spectator platforms that surrounded the court were built in such a way that people could watch certain astronomical events—including equinoxes, solstices, and changes of the seasons—from them.

CONSTRUCTION AND THE STARS

Many Mayan buildings were constructed based on an advanced knowledge of astronomy. El Castillo, also known as the Temple of Kukulcan, at Chichén Itzá in Yucatán, Mexico, is a prime example. In fact, many believe that this building was actually constructed as a three-dimensional solar calendar.

El Castillo has 91 stairs on each side; when added together along with the final step at the top of the platform, there are 365 total steps, which equals the number of days in the solar calendar. The number 91 also represents the number of days between the winter solstice, spring equinox, summer solstice, and fall equinox. The pyramid is about 100 feet (30 m) tall and is made up of eighteen stepped levels. The number eighteen is significant because it represents the eighteen months of the Mayan solar calendar. There are fifty-two panels on each triangular face of the pyramid. These panels represent the number of years in the Mayan Calendar Round. There are two giant carvings of snake heads at the base of the north staircase; these are believed to represent the god Kukulcán. Twice a year, these snake heads present an illusion of turning into an actual snake. During the spring and fall equinoxes, the pyramid's stepped formation forms a shadow in the shape of a snake's body. The building is also structured so that the west side faces the sunset on the day of the zenith passage. A single example of the building's astronomical connections may be considered a coincidence, but there are simply too many obvious design and architectural choices in El Castillo for there to be any doubt that the Maya had sharply focused STEM knowledge, which they brought to bear on their construction projects.

CHAPTER SIX

MAYAN MYSTERIES STILL BEING REVEALED

The Mayan civilization has been a mystery to researchers for centuries. When the Spanish arrived in the Americas, the Maya were still living in Mesoamerica. However, most of them were dwelling in small tribes and villages, some tucked deep in jungles and mountainous regions. The only known city that the Maya were inhabiting when the Spanish arrived was Tulum, a city overlooking the Caribbean coast. All of the other great cities—with their impressive pyramids, magnificent palaces, and intricately designed structures—were already abandoned. In some instances, no one had lived in these urban areas for centuries, and the rain forest had already shrouded them with dense foliage and jungle plants.

SUDDEN DEPARTURE

No one is really certain why the Maya moved away from their grand city-states. Sometime around 900 BCE, most of the cities had been abandoned. The people moved away, possibly migrating to other regions via the sacbeob, and left the cities to become massive ghost towns.

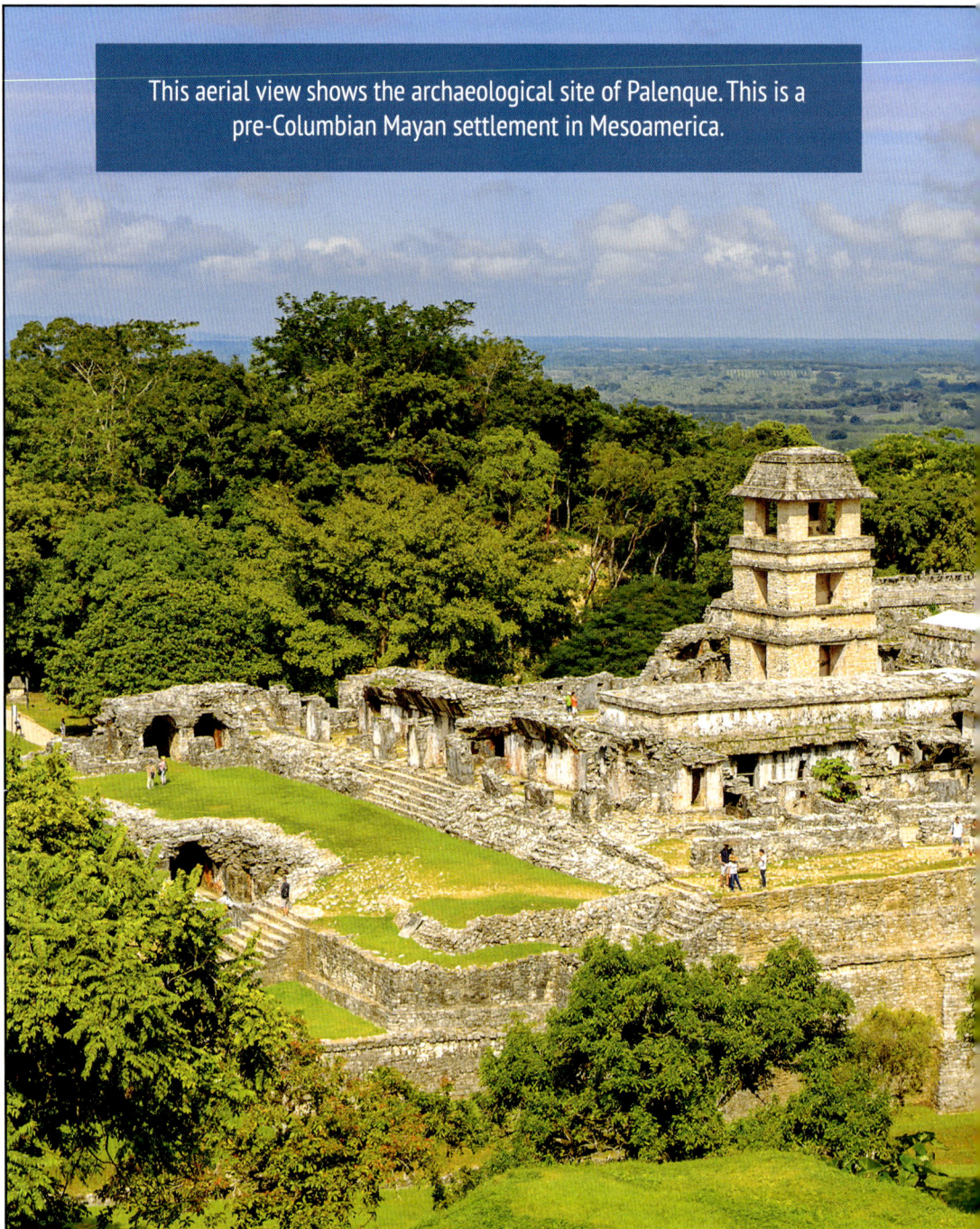

This aerial view shows the archaeological site of Palenque. This is a pre-Columbian Mayan settlement in Mesoamerica.

Some historians theorize that the Maya may have abandoned their cities due to issues associated with drought and famine. It is suspected that the huge populations of this region eventually took a disastrous toll on the environment. Thousands of trees were cut down for farmland, city growth, and industry (i.e., lime plaster production). Tropical rain forest land is also not known for its fertile soil. Therefore, when large swaths of forest are cut down, the land turns to desert over time. With few plants being able to grow on the soil, agricultural production would have decreased and famine could have resulted. As the forests diminished, so, too, would the rainfall. Thus, drought would have been a natural result of deforestation, which would have left the people without the water they needed to survive.

STILL THRIVING TODAY

Regardless of the reasons that the Maya abandoned their cities, it is important to remember that the Maya as a people still do exist. People often get the idea that since the Mayan cities are in ruins, then the Mayan people must have also disappeared. On the contrary, there are millions of Maya living in the Central American region today who are direct descendants of this ancient civilization.

The modern-day Maya still use some of the technology implemented by their ancestors. The people who live in the Mayan highlands,

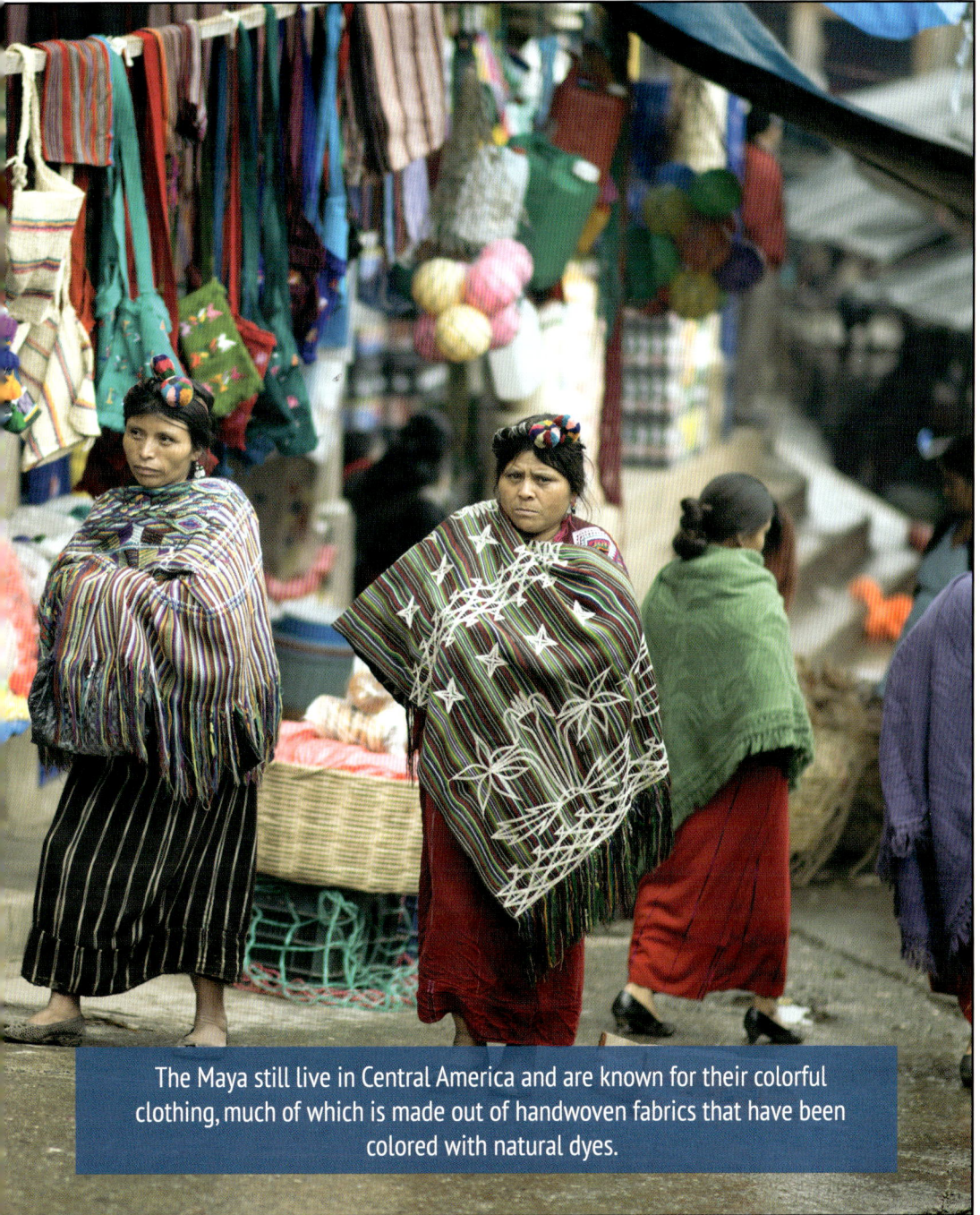

The Maya still live in Central America and are known for their colorful clothing, much of which is made out of handwoven fabrics that have been colored with natural dyes.

for example, use the Mayan calendar for their day counting and religious rituals. They also work in agriculture, just as many of their ancient forefathers did. They use age-old techniques of cutting and burning forestland to obtain the land they need to adequately farm.

HIDDEN IN THE JUNGLE

Many Mayan sites have been discovered and excavated. However, many more are hidden, lost in the dense Central American jungle. When a person looks out over this rain forest landscape, all they can see is a sea of green. What they may not realize, though, is that underneath all of that jungle, enormous Mayan cities might be tucked away, loaded with countless clues about the ancient people who built them.

Today, archaeologists use advanced technology to uncover these ancient sites. One important piece of equipment is lidar. Sometimes spelled as LiDAR (which stands for light detection and ranging), this technology has been used to discover enormous Mayan cities. One example was discovered in Guatemala's highlands in February 2018. This particular city covered more than 800 square miles (2,072 sq km) and had more than sixty thousand structures. The scans also showcased canal systems, roadways, and even defensive walls and fortifications. "There were urban areas, there were rural areas, there were areas in between, there were empty areas. It just looks and reads like the landscape of a complex society," explained Marcello

This picture of dense Central American jungle begs the question: how many ancient Mayan sites lie hidden beneath that sea of green?

Canuto—the director of the Middle American Research Institute at Tulane University—in a *Discover* article by Nathaniel Scharping.

This particular modern piece of technology has totally revolutionized archaeological research in

HOW LIDAR WORKS

Modern lidar equipment is made up of a laser, scanner, and a GPS receiver. With lidar, scientists collect data about objects or areas from a distance. To use lidar, scientists attach the system to the bottom of a plane or helicopter and then send out pulsed lasers over a specific area—typically hundreds of thousands of pulses every second. The laser beams strike the earth and then bounce back to the transmitter. The device measures the time it takes for the light to return to the transmitter and converts that data into a three-dimensional image of whatever it is scanning.

Think about the topography of a particular area. If the area is completely flat, then the light will all return to the device at about the same rate. This lets the lidar construct an image that shows a flat plain. However, if some of the topography is mountainous, then the light that hits the higher points of a mountain will arrive back to the device more quickly than at areas in the valleys. Thus, the resulting image would show the varying highs, lows, and contours of the land.

Lidar is used to examine both natural topography and manmade structures. It has been particularly useful in peering through the dense foliage of the Central American rain forest to reveal the mysteries hidden beneath the jungle's canopy.

Central America. Lidar is allowing scientists to quickly find out what is hidden in the jungle instead of having to take years and years excavating sites that are only suspected to have historical significance. Archaeologists feel confident that, with this technology and any other helpful tools that are invented in the future, they will continue to find out about the mysteries of the amazing Maya. After all, history has only just scratched the surface on learning about this ancient civilization and the many ways it impacted the world through its advanced STEM knowledge.

TIMELINE

2600 BCE The Mayan civilization is believed to begin.

2000 BCE Village farming begins.

700 BCE The Mayan glyph writing system is developed.

500 BCE El Mirador, the oldest-known large Mayan city-state, is established. At its peak, about one hundred thousand people lived in the area.

400 BCE The oldest-known stone Mayan calendars are created.

300 BCE Mayan civilizations are ruled by kings and nobles.

50 BCE The city of Cerros is built, with a complex of temples and ball courts. It is a population center for about one hundred years before it is abandoned for unknown reasons.

150 CE El Mirador is abandoned.

200 CE Tikal is established as a great Mayan city in the rain forests of northern Guatemala. At its height, it becomes one of the largest city-states in Mesoamerica with as many as five hundred thousand residents.

600 CE Chichén Itzá is established in the Yucatán and will remain an important Mayan city until the 1200s.

683 CE   Emperor Pakal dies and is buried at the Temple of the Inscriptions at Palenque.

751 CE   Alliances and trade between Mayan cities begins to disintegrate.

899 CE   Tikal is abandoned.

900 CE   The Classic period of Mayan history ends with the abandonment of the southern lowland cities. However, cities in the Yucatán continue to thrive.

1200 CE   Northern Mayan cities begin to be abandoned.

GLOSSARY

apex The point at which all sides of a pyramidal structure join together.

aqueduct A manmade channel that is meant to allow water to flow from the source to a location farther away.

causeways Roads that have been built at a slightly higher elevation than the soggy ground or water that surrounds them.

cenote A natural sinkhole that is filled with water and deemed of sacred importance to the Maya.

codices Manuscripts that are compiled into a large book format.

corbel arch An arch formed by stones that span a gap using an inverted V formation.

culvert A pipe or drain used to divert water flow.

deforestation The systematic and widespread destruction of woodland areas.

hieroglyphics A form of writing that is made up of symbols, signs, and pictures.

latex A natural secretion of some plants; used in the creation of rubber.

lidar A technology that uses timed laser measurements to create a three-dimensional image of something.

Mesoamerica A region in Central America where many pre-Columbian societies flourished. The countries of Belize, Guatemala, El Salvador, Honduras, Nicaragua, and Costa Rica are part of this region.

obsidian Black volcanic glass.

public works Government projects that build such things as roads and utilities (water, sewer, and electricity) for a community.

quarries Places from which large pieces of stone or smaller bits of rock are cut and harvested for use in building projects elsewhere.

sacbeob A network of paved roads built by the Maya (plural of *sacbe*).

stelae Upright stones or columns that have been engraved with inscriptions.

vigesimal A number system that is based on groups of twenty.

water pressure The force that makes flowing water either strong or weak in an enclosed system, such as a pipe or culvert.

zenith The time when the sun is directly over the observer, which occurs only in the tropical zones near the equator.

FOR MORE INFORMATION

American Anthropological Association (AAA)
2300 Clarendon Boulevard, Suite 1301
Arlington, VA 22201
(703) 528-1902
Website: https://www.americananthro.org
Facebook: @AmericanAnthropologicalAssociation
Instagram and Twitter: @AmericanAnthro
Founded in 1902, this nonprofit organization is
    the world's largest association of professional
    anthropologists. The organization publishes
    more than twenty journals and offers career and
    professional development services, provides
    scholarships for university students, and hosts
    research conferences twice a year.

Ancient History Encyclopedia
Brook House, Mint Street
Godalming, Surrey GU7 1HE
United Kingdom
+44 7956-670133
Website: https://www.ancient.eu
Facebook, Instagram, and Twitter: @ahencyclopedia
The Ancient History Encyclopedia project is dedicated
    to expanding global understanding of history and
    world cultures. Its website has entries and articles
    on many ancient civilizations, including the Maya.

Canadian Archaeological Association (CAA)
Website: https://canadianarchaeology.com
Facebook: @CanadianArchaeologicalAssociation
Twitter: @can_arch
The CAA is made up of professional scholars,
    students, and enthusiasts who are interested
    in archaeology. The organization is committed
    to increasing archaeological knowledge
    across Canada.

Canadian Historical Association (CHA)
1912-130 Albert Street
Ottawa, ON K1P 5G4
Canada
(613) 233-7885
Website: https://cha-sha.ca
Facebook: @CanadianHistoricalAssociation
    -SociétéhistoriqueduCanada
Twitter: @CndHistAssoc
The CHA is an association of professionals and those
    with an interest in historical study. Its goal is to
    promote cultural and scholarly understanding
    through history.

Foundation for the Advancement of Mesoamerican
    Studies, Inc. (FAMSI)
5905 Wilshire Boulevard
Los Angeles, CA 90036
(323) 857-6000
Website: http://www.famsi.org
Facebook: @famsi.org
This foundation's goal is to promote an increased
    understanding of the many Mesoamerican
    cultures—including Mayan. It helps support
    scholars in this field to further their research in
    anthropology, archaeology, art history, and other
    related fields.

National Institute of Culture and History
Government House/House of Culture, Regent Street
Belize City
Belize
+501-227-0811
Website: https://nichbelize.org
Facebook: @NICHBZ
Instagram: @museumofbelize
Twitter: @NICHBelize
This institute's main goal is to preserve and protect
    Belize's culture and history. It hosts numerous
    educational opportunities throughout the year that
    focus on Mayan culture and the ancient Mayan
    ruins located within the country of Belize.

Peabody Museum of Archaeology & Ethnology
11 Divinity Avenue
Cambridge, MA 01138
(617) 496-1638
Website: https://www.peabody.harvard.edu
Facebook and Twitter: @PeabodyMuseum
As one of the oldest archaeological museums in the
world, the Peabody at Harvard houses a variety of
artifacts and art from various world cultures, with
an emphasis on those native to the Americas.

Society for American Archaeology
1111 14th Street NW, Suite 800
Washington, DC 20005-5622
(202) 789-8200
Website: https://www.saa.org
Facebook: @SAAorgfb
Instagram: @societyforamericanarchaeology
Twitter: @SAAorg
This organization's mission is to help the general
public gain a better understanding of and
appreciation for humanity's past. The group
promotes research, encourages the investigation
into archaeological records, works to protect
artifacts and dig sites, offers educational
opportunities for the public, and publishes
research findings.

FOR FURTHER READING

Coe, Michael D. *The Maya*. London, UK: Thames & Hudson, 2015.

Doughtery, Martin J. *The Aztec, Inca, and Maya Empires: The Illustrated History of the Ancient Peoples of Mesoamerica and South America*. London, UK: Amber Books, 2018.

Honders, Christine. *Ancient Maya Culture*. New York, NY: PowerKids Press, 2017.

Hofer, Charles. *Ancient Maya Technology*. New York, NY: PowerKids Press, 2017.

Hunter, Nick. *Daily Life in the Maya Civilization*. London, UK: Raintree, 2016.

Mahoney, Emily. *The Mysterious Maya Civilization*. New York, NY: Greenhaven, 2018.

Phillips, Charles. *The Complete Illustrated History of the Aztec & Maya: The Definitive Chronicle of the Ancient Peoples of Central America and Mexico Including the Aztec, Maya, Olmec, Mixtec, Toltec and Zapotec*. Irvine, CA: Hermes House, 2015.

Williams, Brian, and Caroline Dodds Pennock. *Maya, Incas, and Aztecs*. New York, NY: DK Publishing, 2018.

Witschey, Walter R. T. *Encyclopedia of the Ancient Maya*. Lanham, MD: Rowman & Littlefield, 2016.

BIBLIOGRAPHY

Atlas Obscura. "La Danta." Retrieved February 25, 2019. https://www.atlasobscura.com/places /la-danta.

Babowice, Hope. "How the Maya Built Their Pyramids, Buildings." May 2, 2016. https://www.dailyherald .com/article/20160502/submitted/160509869.

Bartlett, Ray. "Exploring El Mirador, Guatemala's Mysterious Mayan Ruin." Lonely Planet, November 2018. https://www.lonelyplanet.com/guatemala /travel-tips-and-articles/exploring-el-mirador -guatemalas-mysterious-mayan-ruin/40625c8c -8a11-5710-a052-1479d2757a75.

Brown, Chip. "El Mirador, the Lost City." *Smithsonian*, May 2011. https://www.smithsonianmag.com /history/el-mirador-the-lost-city-of-the-maya -1741461.

Choi, Charles Q. "Ancient Mayans Likely Had Fountains and Toilets." Live Science, December 23, 2009. https://www.livescience.com/5959-ancient-mayan -fountains-toilets.html.

DK Find Out! "Temple of the Inscriptions." DK Find Out! Retrieved February 26, 2019. https://www .dkfindout.com/us/history/mayans/temple -inscriptions.

Fash, William. *The Maya: Death of an Empire*. Directed by Dana K. Ross. New York, NY: A&E Television Networks, 2008.

French, Kirk D. "Palenque's Water Management."
Pre-Columbian Art Research Institute. Retrieved
February 20, 2019. http://www.precolumbia.org
/pari/palenque/dig/report/mapping/media/Water
_Management.pdf.

Guarino, Ben. "This Major Discovery Upends Long-Held
Theories about the Maya Civilization." *Washington
Post*, September 27, 2018. https://www
.washingtonpost.com/science/2018/09/27
/this-major-discovery-upends-long-held-theories
-about-maya-civilization/?noredirect=on&utm
_term=.10c0177f836c.

History.com. "Maya." History.com. Retrieved February
24, 2019. https://www.history.com/topics/ancient
-americas/maya.

Jarus, Owen. "Early Urban Planning: Ancient Mayan
City Built on Grid." Live Science, April 29, 2015.
https://www.livescience.com/50659-early-mayan
-city-mapped.html.

Logan, Jim. "An Ancient Mayan Copernicus." Current,
August 16, 2016. https://www.news.ucsb
.edu/2016/017062/mayan-moment.

Lorenzi, Roseela. "Ancient Mayan Superhighways
Found in the Guatemala Jungle." Seeker, January
27, 2017. https://www.seeker.com/ancient
-mayan-superhighways-found-in-the-guatemala
-jungle-2219303581.html.

Moskowitz, Clara. "Secret Myan Blue Paint Found." Live Science, February 25, 2008. https://www.livescience.com/2322-secret-mayan-blue-paint.html.

Penn State. "Maya Plumbing: First Pressurized Water Feature Found in New World." ScienceDaily, May 5, 2010. www.sciencedaily.com/releases/2010/05/100504155421.htm.

Prostak, Sergio. "Largest Ancient Maya Dam Found in Guatemala." SciNews, July 17, 2012. http://www.sci-news.com/archaeology/article00470.html.

Scharping, Nathaniel. "LiDAR Scans Reveal Maya Were Far Bigger and More Complex than Thought." Discover, September 27, 2018. http://blogs.discovermagazine.com/d-brief/2018/09/27/maya-lidar-scans-60000-new-structures/#.Xlc4U4hKiUk.

Slivka, Kelly. "A Mayan Water System with Lessons for Today." New York Times, July 16, 2012. https://green.blogs.nytimes.com/2012/07/16/a-mayan-water-system-with-lessons-for-today.

Wade, Lizzie. "The City at the Beginning of the World." Archaeology, July/August 2018. https://www.archaeology.org/issues/303-1807/features/6684-maya-urban-grid.

Zapp, Ivar. Atlantis in America: Navigators of the Ancient World. Kempton, IL: Adventures Unlimited Press, 1998.

INDEX

ABOUT THE AUTHOR

Amie Jane Leavitt graduated from Brigham Young University and is an accomplished author, researcher, and photographer. She has written numerous books for young readers, has contributed to online and print media, and has worked as a consultant, writer, and editor for educational publishing and assessment companies.

PHOTO CREDITS

Cover Brit Finucci/Moment/Getty Images; p. 5 Peter Hermes Furian/Shutterstock.com; pp. 8–9 DC_Aperture/Shutterstock.com; pp. 10–11 Peter Batarseh/Shutterstock.com; p. 14 Yummyphotos/Shutterstock.com; p. 20 Dmitry Rukhlenko/Shutterstock.com; p. 22 SherSS/Shutterstock.com; p 27 DEA/M. Borchi/De Agostini/Getty Images; p. 31 Mattia Marenco/Alamy Stock Photo; p. 34 Leon Rafael/Shutterstock.com; pp. 36–37 chrisontour84/Shutterstock.com; p. 41 Estuardo Ajin/Shutterstock.com; p. 43 Alexander Ryabintsev/Shutterstock.com; p. 45 Ikpro/Shutterstrock.com; p. 49 Chris Hill/Shutterstock.com; p. 51 Robert Michael/AFP/Getty Images; p. 52 Raif Broskvar/Shutterstock.com; pp. 56–57 Anton_Ivanov/Shutterstock.com; pp. 58–59 Ray Waddington/Alamy Stock Photo; pp. 60–61 Wollertz/Shutterstock.com; cover and interior pages (dark textured background) Midiwaves/Shutterstock.com; interior pages (scroll pattern page borders) Megin/Shutterstock.com, (yellow marbled page borders) Chizhovao/Shutterstock.com.

Design and Layout: Nicole Russo-Duca; Editor: Siyavush Saidian; Photo Researcher: Nicole Baker